European Federation of Corrosion Publications

NUMBER 16
Third Edition

A Working Party Report on

Guidelines on Materials Requirements for Carbon and Low Alloy Steels For H_2S-Containing Environments in Oil and Gas Production

The Institute of Materials, Minerals & Mining

Published for the European Federation of Corrosion by Maney Publishing on behalf of The Institute of Materials, Minerals & Mining

Published by Maney Publishing for the European Federation of Corrosion, on behalf of The Institute of Materials, Minerals & Mining

Maney Publishing is the trading name of W.S. Maney & Son Ltd.

Maney Publishing, Suite 1C, Joseph's Well, Hanover Walk, Leeds LS3 1AB, UK

First published 2009 by Maney Publishing

Maney Publishing ISBN-13: 978-1-90654-033-3 (book)
Maney Publishing stock code: B808
ISSN 1354-5116

The publisher's policy is to use permanent paper from mills that operate a sustainable forestry policy, and which has been manufactured from pulp which is processed using acid-free and elementary chlorine-free practices.

Typeset and printed by Charlesworth Group

Contents

European Federation of Corrosion (EFC) publications:
Series introduction

The EFC, incorporated in Belgium, was founded in 1955 with the purpose of promoting European co-operation in the fields of research into corrosion and corrosion prevention.

Membership of the EFC is based upon participation by corrosion societies and committees in technical Working Parties. Member societies appoint delegates to Working Parties, whose membership is expanded by personal corresponding membership.

The activities of the Working Parties cover corrosion topics associated with inhibition, education, reinforcement in concrete, microbial effects, hot gases and combustion products, environment-sensitive fracture, drinking water environments, marine environments, refineries, surface science, physico-chemical methods of measurement, the nuclear industry, the automotive industry, computer-based information systems, coatings, tribo-corrosion, polymeric materials and the oil and gas industry. Working Parties and Task Forces on other topics are established as required.

The Working Parties function in various ways, e.g. by preparing reports, organising symposia, conducting intensive courses and producing instructional material, including films. The activities of Working Parties are co-ordinated, through a Science and Technology Advisory Committee, by the Scientific Secretary. The administration of the EFC is handled by three Secretariats: DECHEMA e.V. in Germany, Fédération Française pour les sciences de la Chimie in France, and the Institute of Materials, Minerals and Mining in the UK. These three Secretariats meet at the Board of Administrators of the EFC. There is an annual General Assembly at which delegates from all member societies meet to determine and approve EFC policy. News of EFC activities, forthcoming conferences, courses, etc., is published in a range of accredited corrosion and certain journals throughout Europe. More detailed descriptions of activities are given in a Newsletter prepared by the Scientific Secretary.

The output of the EFC takes various forms. Papers on particular topics, e.g. reviews or results of experimental work, may be published in scientific and technical journals in one or more countries in Europe. Conference proceedings are often published by the organisation responsible for the conference.

In 1987 the then Institute of Metals was appointed as the official EFC publisher. Although the arrangement is non-exclusive and other routes for publication are still available, it is expected that the Working Parties of the EFC will use The Institute of Materials, Minerals and Mining for publication of reports, proceedings, etc., wherever possible.

The name of The Institute of Metals was changed to The Institute of Materials on 1 January 1992 and to the Institute of Materials, Minerals and Mining with effect from 26 June 2002. The series is now published by Maney Publishing on behalf of the Institute of Materials, Minerals and Mining.

P. McIntyre
EFC Series Editor, the Institute of Materials, Minerals and Mining, London, UK

EFC Secretariats are located at:

Dr B. A. Rickinson
European Federation of Corrosion, the Institute of Materials, Minerals and Mining, 1 Carlton House Terrace, London, SW1Y 5DB, UK

Dr J. P. Berge
Fédération Européenne de la Corrosion, Fédération Française pour les sciences de la Chimie, 28 rue Saint-Dominique, F-75007 Paris, FRANCE

Professor Dr G. Kreysa
Europäische Föderation Korrosion, DECHEMA e. V., Theodor-Heuss Allee 25, D-60486, Frankfurt, GERMANY

Preface

The presence of H_2S in oil and gas production poses its own specific threat to the integrity of the production system Many materials suffer from cracking of various forms when exposed to H_2S which may result in the catastrophic failure of equipment with the attendant risk of releasing the contents into the environment. Besides the general risks associated with release of hydrocarbons in terms of pollution and fire, the release of H_2S exposes persons in the vicinity to the risks of poisoning and death.

For these reasons the materials engineer is mindful of the need to select materials of proven resistance to cracking in H_2S-containing environments.

This guideline document is specifically concerned with the material requirements for carbon and low alloy steels for H_2S-containing oil and gas field service. It aims to be comprehensive in considering all possible types of cracking which may result from exposure of such steels to H_2S, the conditions under which they may occur and appropriate materials requirements to prevent such cracks. In addition, the document recommends test methods for evaluating materials performance and particularly focuses on a fitness-for-purpose approach whereby the test conditions are selected to reflect the realistic service conditions.

Thus, this guideline document is believed to be a practical, industry-oriented guide to the subject. It incorporates much of the recent developments in knowledge on the way in which the detailed environmental conditions affects risk of cracking. It also recognises conditions in which some relaxation of strict requirements may be made which can result in considerable cost saving without any increase in risk. Furthermore, it is believed to be the first document which tackles, in one volume, all the H_2S-related cracking problems of all items of equipment used in the oilfield – from the well to the export pipelines.

It is hoped that this guideline document will prove to be a key reference document for materials engineers and product suppliers working in the oil and gas industry.

Svein Eliassen
Chairman (1993-1998)
Carbon and Low Alloy Steels
Working Group of the Working Party
European Federation of Corrosion

Liane Smith
Chairman (1993–1998)
Working Party on Corrosion
in Oil and Gas Production
European Federation of Corrosion

Second Edition Note

After the first publication of EFC16 in 1995, two joint industry sponsored projects were established to investigate safe hardness limits for welds in carbon and low alloy steels in H_2S-containing environments.[1,2] This edition incorporates the results of those projects, following the guidance of ISO 15156, in section 8.2.1., Table 8.1. Other changes to the text are mostly editorial.

Liane Smith
Chairman (1998–2001)
Carbon and Low Alloy Steels
Working Group of the Working Party
European Federation of Corrosion

Phil Jackman
Chairman (1998–2001)
Working Party on Corrosion
in Oil and Gas Production
European Federation of Corrosion

Third Edition Note

Since the previous edition, experience has highlighted improvements which could be made to the guidance given in this document, particularly for test methods. Specifically it was found that the test solution was difficult to maintain within the pH range over the full time. Frequent adjustments to the pH resulted in the risk of air entering the test chamber. Other aspects which lacked clarity concerned the evaluation of HIC and C-ring and four-point bend test samples. Improvements to the text on these matters, plus minor editorial improvements, are incorporated in this third edition.

Liane Smith
Chairman (1998–2008)
Carbon and Low Alloy Steels
Working Group of the Working Party
European Federation of Corrosion

Stein Olsen
Chairman (2001–2008)
Working Party on Corrosion
in Oil and Gas Production
European Federation of Corrosion

1

Definitions, Abbreviations and Symbols

Acidising : Well treatment using acid, usually to improve well production rate.

Blistering : *See* SWC.

Bubble Point Pressure : The pressure under which gas bubbles will form in a liquid at a particular operating temperature.

Carbon Steel : An alloy of carbon and iron containing up to 0.8% carbon and up to 1.65% manganese and residual quantities of other elements, except those intentionally added in specific quantities for deoxidation (usually silicon and/or aluminium).

Cementite : A microstructural constitutent of steels composed principally of iron carbide.

Cold Forging : *See* Cold Working.

Cold Reducing : *See* Cold Working.

Cold Working : Deforming metal plastically under conditions of temperature and strain rate that induce strain hardening, usually, but not necessarily, conducted at room temperature. Contrast with hot working.

CLR* : Crack Length Ratio.

CTR* : Crack Thickness Ratio.

CR : C-ring testing.

CSR* : Crack Surface Ratio (also referred to in some documents as Crack Sensitivity Ratio).

DCB : Double cantilever beam testing.

* For definitions of these parameters see NACE TM0284.

ε_{air} : Strain to failure in air.

ε_n : Normalised strain to failure = $\varepsilon_s/\varepsilon_{air}$.

ε_s : Strain to failure in the solution.

EI : Embrittlement index = $1 - RA_s/RA_{air}$.

Fabrication : Metal joining by the use of welding processes.

Ferrite : A body-centred cubic crystalline phase of iron base alloys.

Ferritic Steel : A steel whose microstructure at room temperature consists predominantly of ferrite.

Ferrous Metal : A metal in which the major constituent is iron.

Fitness-For-Purpose : Suitability for use under the expected service conditions.

FPB : Four point bend testing.

Free-Machining Steel : Steel to which elements such as sulphur, selenium, or lead have been added intentionally to improve machinability.

Hardness : Resistance of metal to plastic deformation, usually by indentation.

Heat Treatment : Heating and cooling a solid metal or alloy in such a way as to obtain desired properties. Heating for the sole purpose of hot working is not considered heat treatment.

Heat-Affected Zone (HAZ) : That portion of the base metal that was not melted during brazing, cutting, or welding, but whose microstructure and properties were altered by the heat of these processes.

HIC : Hydrogen induced cracking – also called stepwise cracking (SWC).

HSLA : High Strength Low Alloy Steel.

Hot Rolling : Hot working a metal through dies or rolls to obtain a desired shape.

HPIC : Hydrogen pressure induced cracking — also called stepwise cracking (SWC).

Hot Working : Deforming metal plastically at such a temperature and strain rate that recrystallisation takes place simultaneously with the deformation, thus avoiding any strain hardening.

Internal Cracking : see SWC.

K_{1SSC} : Threshold stress intensity factor above which sulphide stress crack propagation is sustained.

Low-Alloy Steel : Steel with a total alloying element content of less than about 5%, but more than specified for carbon steel. (See also micro-alloyed steel).

Martensite : A supersaturated solid solution of carbon in iron producing a body-centred tetragonal crystalline phase characterised by an acicular (needle-like) microstructure.

Martensitic Steel : A steel in which a microstructure of martensite can be attained by quenching at a cooling rate fast enough to avoid the formation of other microstructures.

Microstructure : The structure of a metal as revealed by microscopic examination of a suitably prepared specimen.

Microalloyed Steel : Steels which contain small additions of carbide and/or nitride forming elements, principally Nb, V, Ti.

MPI : Magnetic Particle Inspection.

OCTG : Oil country tubular goods, i.e. casing and tubing.

Partial Pressure : Ideally, in a mixture of gases, each component exerts the pressure it would exert if present alone at the same temperature in the total volume occupied by the mixture. The partial pressure of each component is equal to the total pressure

		multiplied by its mole or volume fraction in the mixture.
Plastic Deformation	:	Permanent deformation caused by stressing beyond the limit of elasticity.
Pearlite	:	A microstructural constituent of carbon and low alloy steels consisting of alternate lamellae of ferrite and cementite.
Premium Connection	:	A threaded joint for tubular components using a proprietary thread geometry as opposed to an industry standard thread form.
PWHT	:	Post-Welding Heat Treatment, i.e. heating and cooling a weldment in such a way as to obtain desired properties.
Ranking	:	Comparing the relative performance of several materials to establish an order.
R_a	:	Measurement of surface roughness. Arithmetic mean of departure of the roughness profile from the mean line.
RA_{air}	:	Reduction in area in air.
RA_s	:	Reduction in area in solution.
Screening	:	Preliminary evaluation to establish potential materials for detailed testing.
Shape Control	:	A treatment in which inclusions, notably sulphides, are prevented from elongation during hot reduction. This is normally achieved by additions of calcium or rare earth metals.
SOHIC	:	Stress-oriented hydrogen-induced cracking. Staggered small cracks formed approximately perpendicular to the principal stress (residual or applied) resulting in a "ladder-like" crack array linking (sometimes small) pre-existing HIC cracks. Note: The mode of cracking can be categorized as SSC caused by a combination of external stress and the local strain around hydrogen-induced

		cracks. SOHIC is related to SSC and HIC/SWC. It has been observed in parent material of longitudinally welded pipe and in the heat-affected zone (HAZ) of welds in pressure vessels. SOHIC is a relatively uncommon phenomenon usually associated with low-strength ferritic pipe and pressure vessel steels.
SMYS	:	Specified minimum yield strength.
Spheroidise	:	*See* Shape Control.
SSC	:	Sulphide Stress Cracking, i.e. cracking under the combined action of tensile stress and corrosion in the presence of water and hydrogen sulphide.
SSR	:	Slow strain rate testing.
SWC	:	Stepwise cracking. Blistering and cracking principally parallel to the rolling plane of the steel plate.
SZC	:	Soft zone cracking. Form of SSC that may occur when a steel contains a local "soft zone" of low yield strength material. Note: Under service loads, soft zones may yield and accumulate plastic strain locally, increasing the SSC-susceptibility to cracking of an otherwise SSC-resistant material. Such soft zones are typically associated with welds in carbon steels.
σ_{th}	:	Threshold stress.
Tempering	:	In heat treatment, reheating hardened steel or hardened cast iron to, some temperature below the lower critical temperature for the purpose of decreasing the hardness and increasing the toughness. The process is also sometimes applied to normalised steel.
Tensile Strength	:	In tensile testing the ratio of maximum load to original cross-sectional area (reference ASTM A370). Also called 'ultimate strength'.
Tensile Stress	:	The net tensile component of all combined stresses–axial or longitudinal, circumferential or 'hoop', and residual.

UT	:	Uniaxial tensile testing.
Yield Strength	:	The stress at which a material exhibits a specified deviation from the proportionality of stress to strain. The deviation is expressed in terms of strain by either the permanent offset method (usually at a strain of 0.2%) or the total-extension-under-load method (usually at a strain of 0.5%) (reference ASTM A370).

2

Standards Referred to in this Document

API 5CT	American Petroleum Institute, Specification for Casing and Tubing (US Customary Units).
ASTM A370	Standard Test Methods and Definitions for Mechanical Testing of Steel Products.
ASTM A833	Practice for Indentation Hardness of Metallic Materials by Comparison Hardness Testers.
ASTM E10	Test Method for Brinell Hardness of Metallic Materials.
ASTM E18	Test Method for Rockwell Hardness and Rockwell Superficial Hardness of Metallic Material.
ASTM E92	Test Method for Vickers Hardness of Metallic Materials.
ASTM E140	Hardness Conversion Tables for Metals (Relationships between Brinell Hardness, Vickers Hardness, Rockwell Hardness, Rockwell Superficial Hardness and Knoop Hardness).
ASTM E384	Test Method for Microhardness of Material.
ASTM G38	Practice for Making and Using C-Ring Stress Corrosion Test Specimens.
ASTM G39	Practice for Preparation and Use of Bent-Beam Stress Corrosion Test Specimens.
ASTM G49	Standard Practice for the Preparation and Use of Direct Tension Stress Corrosion Test Specimens.
BS 240	Method for Brinell Hardness Test and for Verification of Brinell Hardness Testing Machines.
BS 427	AMD 1. Method for Vickers Hardness Test and for Verification of Vickers Hardness Testing Machines (AMD 6756).

BS 860	Tables for Comparison of Hardness Scales.
BS 891	Rockwell Hardness Test Methods for Hardness Test (Rockwell Method) and for Verification of Hardness Testing Machines (Rockwell Method).
BS 4515	Specification for Welding of Steel Pipelines on Land and Offshore (Part 1: Carbon and Carbon Manganese Steel Pipelines).
DIN 50103-3	Testing of Metallic Materials - Rockwell Hardness Test Part 3: Modified Rockwell Scales Bm and Fm for Thin Steel Sheet.
DIN 50133	Testing of Metallic Materials; Vickers Hardness Test HV 0.2 to HV 100.
DIN 50351	Testing of Metallic Materials; Brinell Hardness Test.
ISO 6506-1	Metallic Materials - Hardness Test - Brinell Test.
ISO 6507 1-3	Metallic Materials - Hardness Test - Vickers Test Part 1: HV 5 to HV 100 Part 2: HV 0.2 to less than HV 5 Part 3: Less than HV 0.2.
ISO 6508	Metallic Materials - Hardness Test - Rockwell Test (Scales A-B-C-D-E-F-G-H-K) (replaces R80 and 2173).
ISO 7539	Corrosion of Metals and Alloys - Stress Corrosion Testing.
ISO 15156	Petroleum and natural gas industries - Materials for use in H_2S containing environments in oil and gas production.
NACE TM0177	Laboratory Testing of Metals for Resistance to Sulfide Stress Cracking and Stress Corrosion Cracking in H_2S Environments.
NACE TM0284	Evaluation of Pipeline and Pressure Vessel Steels for Resistance to Hydrogen-Induced Cracking.

3

Introduction

Carbon and low alloy steels may be susceptible to cracking when exposed to corrosive H_2S-containing environments, usually referred to as sour service.

Failures of various items of H_2S-containing production equipment and pipelines, by various cracking mechanisms, have led to an awareness of the need to set requirements for carbon and low alloy steels when exposed to corrosive conditions containing H_2S.

Guidance from this document has been incorporated into ISO 15156 "Petroleum and natural gas industries – Materials for use in H_2S containing environments in oil and gas production".

4

Scope

This document provides guidelines on the materials requirements for the safe application of carbon and low alloy steels typically used in H_2S-containing environments in oil and gas production systems. It does not include refinery operations.

A description is given of the types of cracking which can arise in wet H_2S-containing environments.

A systematic approach to the definition of sour service in relation to sulphide stress cracking is presented for Oil Country Tubular Goods up to strength level P110, i.e. high strength low alloy steels with homogeneous microstructure. The same methodology can be applied to other grades of steels and welded steels and these may follow similar trends, however, their precise environmental limits are still being established.

Guidelines are given on materials requirements to prevent sulphide stress cracking, stepwise cracking, stress oriented hydrogen induced cracking and soft zone cracking in steels in various product forms. The guidelines do not include requirements for avoiding other forms of corrosion, hydrogen embrittlement or other forms of cracking (e.g. stress corrosion cracking, corrosion fatigue, etc.) that can occur in the absence of H_2S.

Guidelines for materials requirements for corrosion resistant alloys in H_2S-containing environments are covered in EFC17.[3]

Comprehensive guidance is given on procedures for sulphide stress cracking and stepwise cracking testing including suggested acceptance criteria for the various test methods.

Graphs are included for the approximate determination of pH in producing systems.

Recommendations are made on appropriate methods for hardness testing of components and weld zones for service in H_2S-containing environments.

5

Objective

The aims of this document are:

(i) To define the types of cracking and the conditions under which each can occur in carbon and low alloy steels in wet H_2S-containing environments.
(ii) To provide guidelines on the materials requirements necessary to prevent such cracking (see Section 8).
(iii) To provide procedures and guidelines for test methods for evaluating materials performance (see Annexes A and B).

Corrosive conditions may lead to failures by mechanisms other than hydrogen assisted cracking and should be mitigated by other corrosion control measures which are outside the scope of this document.

6

Types of Cracking in Wet H_2S-Containing Environments

6.1. General

It is characteristic of corrosion in wet H_2S that atomic hydrogen, resulting from an electrochemical reaction between the metal and the H_2S-containing medium, enters the steel at the corroding surface.

The presence of hydrogen in the steel may, depending upon the type of steel, the microstructure and inclusion distribution, and the tensile stress distribution (applied and residual) cause embrittlement and possibly cracking. A brief description of the three main types of cracking is given in the following sections.

6.2. Sulphide Stress Cracking (SSC)

This type of cracking occurs when atomic hydrogen diffuses into the metal but remains in solid solution in the crystal lattice. This reduces the ductility and deformability of the metal which is termed hydrogen embrittlement. Under tensile stress, whether applied or residual from cold-forming or welding etc., this embrittled metal readily cracks to form sulphide stress cracks. The cracking process is very rapid and has been known to take as little as a few hours for a crack to form and cause catastrophic failure.

The tendency for SSC to occur is increased by the presence of hard microstructures such as untempered or partly tempered low temperature transformation products (martensite, bainite). These microstructures may be inherently present in high strength low alloy steels or may result form inadequate or incorrect heat treatment. Hard microstructures may also arise in welds and particularly in low heat input welds in the heat affected zones. Control of hardness, within the limits given in Section 8, has been found to correlate with prevention of SSC in sour environments.

6.3. Stepwise Cracking (SWC)

The name "stepwise cracking" is given to surface blistering and cracking parallel to the rolling plane of the steel plate which may arise without any externally applied or residual stress. The terms used to define such cracking include:

- Blistering,
- Internal cracking,
- Stepwise cracking (SWC),
- Hydrogen-induced cracking (HIC),
- Hydrogen pressure induced cracking (HPIC).

Such cracks occur when atomic hydrogen diffuses in the metal and then recombines as hydrogen molecules at trap sites in the steel matrix. Favourable trap sites are typically found in rolled products along elongated inclusions or segregated bands of microstructure.

The molecular hydrogen is trapped within the metal at interfaces between the inclusions and the matrix and in microscopic voids, with first a crack initiation phase and then propagation along the metallurgical structures sensitive to this type of hydrogen embrittlement.

As more hydrogen enters the voids the pressure rises, deforming the surrounding steel so that blisters may become visible at the surface. The steel around the crack becomes highly strained and this can cause linking of adjacent cracks to form SWC. The arrays of cracks have a characteristic stepped appearance.

Whilst individual small blisters or hydrogen induced cracks do not affect the load bearing capacity of equipment they are an indication of a cracking problem which may continue to develop unless the corrosion is stopped. At the stage when cracks link up to form SWC damage these may seriously affect the integrity of equipment. Failures due to these types of cracking have arisen within months of start-up, whilst crack growth and linking may sometimes take years depending upon the severity of the environment and the susceptibility of the steel. Control of the microstructure, and particularly the cleanliness of steels, as described in Section 8.3 reduces the availability of crack initiation sites and is therefore critical to the control of SWC.

6.4. Stress Oriented Hydrogen Induced Cracking (SOHIC)/Soft Zone Cracking (SZC)

SOHIC and SZC are related to both SSC and SWC. In SOHIC staggered small cracks are formed approximately perpendicular to the principal stress (applied or residual) resulting in a "ladder-like" crack array. The mode of cracking can be categorised as SSC caused by a combination of external stress and the local straining around hydrogen induced cracks. SOHIC has been observed in parent material of longitudinally welded pipe.

Soft Zone Cracking is the name given to a similar phenomenon when it occurs specifically in softened heat affected zones of welds in rolled plate steels. The susceptibility of such weld regions to this type of cracking is thought to arise because of a combination of microstructural effects caused by the temperature cycling during welding and local softening in the intercritical temperature heat affected zone. This results in strains within a narrow zone which may approach or even exceed the yield strain.

SOHIC has caused service failures of pipelines in the past[4, 5] but there are no reported service failures by SOHIC in modern microalloyed line pipe steels produced for service in H_2S with mandatory testing for SWC and SSC.

7

Environmental Factors Affecting Cracking in H_2S-Containing Environments

7.1. General

The cracking mechanisms considered in this document all result from corrosion in the presence of H_2S followed by hydrogen-uptake in the steel. For each of the cracking mechanisms there is a critical hydrogen uptake rate and/or hydrogen concentration in the steel below which cracking does not initiate. The hydrogen uptake is dependent upon a number of parameters of which the most important are:

- hydrogen sulphide concentration;
- pH; and
- temperature.

Other parameters such as CO_2 content, water content and composition, flow rates, surface condition (rust, mill scale, corrosion layers, etc.) and presence of corrosion inhibitors, etc. may have direct or indirect influence on hydrogen uptake and therefore on the risk of cracking.

In the case of SSC and SOHIC, cracking is also controlled by the applied stress (including the effect of total pressure of the system) and residual stresses from working, forming or welding operations.

For practical purposes, and based on field experience, the environmental parameters affecting SSC can be simplified to H_2S concentration and pH (Section 7.2). Environments in which sulphide stress cracking can occur are referred to as sour.

For the other types of cracking (SWC and SOHIC), a general guideline to limiting environmental parameters below which cracking does not occur is difficult to provide because these types of cracks are very much dependent on the internal steel quality, i.e. number and types of inclusions, microsegregations etc. If the steel quality is less than desirable, even traces of H_2S can cause cracking. Thus, if H_2S is present, even at trace levels, the guidelines of Sections 8.3 and 8.4 should be followed to reduce the risk of cracking due to SWC or SOHIC respectively.

7.2. Environmental conditions influencing SSC

7.2.1. Sulphide stress corrosion cracking domains

The occurrence of SSC depends on both pH and H_2S partial pressure. Figure 7.1 represents the general guideline for the occurrence of SSC in terms of pH and H_2S partial pressure.

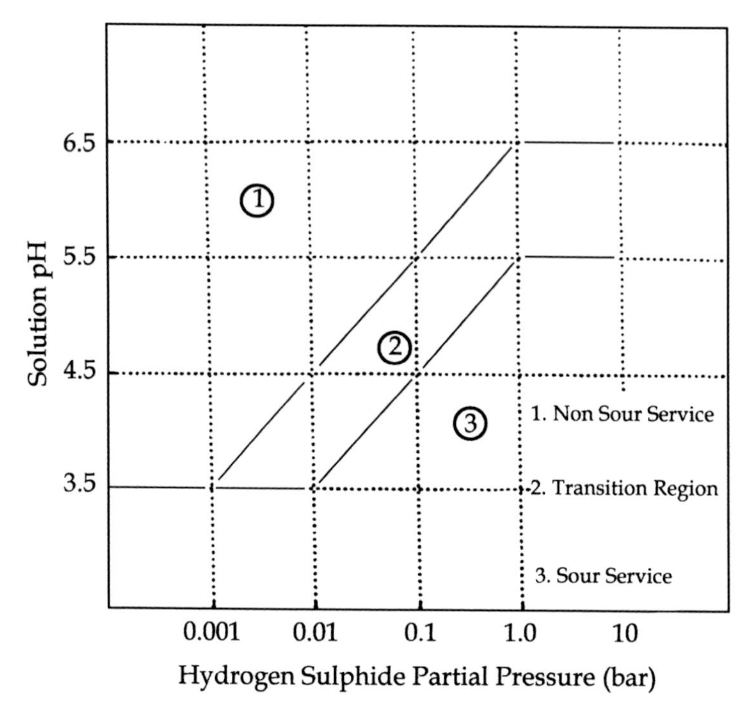

Fig. 7.1 *Sulphide stress corrosion cracking domains as a function of pH and hydrogen sulphide partial pressure.[6] Note: These domains are not applicable to SWC and SOHIC.*

Figure 7.1 has been established for OCTG materials up to grade P110.[6] Other materials of higher strength or with inhomogeneous regions (e.g. welds) are expected to follow a similar trend but the precise positions of the boundary lines of the various regions may be different and have yet to be established.

Thus, in Fig. 7.1 **Region 1** is the domain where SSC did not occur. Steels used in this domain up to strength P110 may be exposed to the condition described by this domain without meeting the metallurgical requirements of Section 8.2 of this document.

Region 2 is a transition zone within which some judgement has to be made regarding the critical metallurgical requirements. In some circumstances, materials with less stringent metallurgical requirements or controls may be acceptable for use provided they meet fitness-for-purpose criteria (for example by testing according to Annex A). For applications where there would be no cost benefits in reduced metallurgical requirements region 3 should be assumed to include region 2.

Region 3 is the domain where SSC may occur. Steels used in this domain should be selected to be resistant to SSC according to the requirements of Section 8.2 of this document. Note that below pH 3.5 the threshold level of H_2S below which SSC does not occur is so low that although it could be quantified in the laboratory the values obtained could not be practically applied in field conditions.

For environments, materials or welded items which are not adequately covered by Fig.7.1, the resistance to SSC can be established by following the test methods described in Annex A. Guidance on requirements for welded steel is given in Section 8.2.1.

Determination of H_2S partial pressure

For gas systems, the partial pressure of H_2S equals the mole fraction of H_2S multiplied by the total system pressure.

For liquid systems (for which no equilibrium gas composition is available), the effective thermodynamic activity of H_2S is defined by a virtual partial pressure of H_2S which may be determined in the following way.

- Determine the bubble point pressure (Bp) of the fluid at operating temperature.
- Determine the mole fraction of H_2S (moles H_2S) in the gas phase at bubble point conditions.
- Calculate the partial pressure of H_2S in the gas at the bubble point (Bp × moles H_2S).
- Use this as the hydrogen sulphide partial pressure value in Fig. 7.1.

Determination of pH

The determination of pH in the actual operating conditions is described.[7,8]

A general guide for an approximate determination of pH for various conditions, if a proven calculation or reliable *in situ* measuring technique is not available is given in Annex C and Ref. [9]. The deviation may be taken as +0.5 and −0.0 pH unit for Safety.

It is important to be aware that excursions to low pH outside of normal operations (e.g. during acidising of wells) may expose equipment to conditions which could give rise to SSC.

In the absence of any data from which the pH can be reliably derived, it is advised to assume that the pH is 3.5 or less.

7.2.2. Influence of Temperature

The embrittling effect of hydrogen is reduced at higher temperatures. This benefit of higher temperatures can be used to relax requirements for materials which are continuously exposed to minimum temperatures in excess of 65°C. This would not apply to facilities which may cool down during normal operations (e.g. shut-downs etc.) but it would apply to tubing, casing and other downhole equipment if these would not be exposed to temperatures below 65°C during service (see Section 8.2.1).

8

Guidelines to Avoid Cracking

8.1. General

Where service will require exposure to high H_2S partial pressures, equipment which has had extensive air contact and has become rusty should be treated with corrosion inhibiting fluids before coming into contact with H_2S.

8.2. Guidelines to Avoid SSC

Materials which meet the requirements given in 8.2.1, 8.2.2 and 8.2.3 are expected to have satisfactory resistance to SSC in regions 2 and 3 of Fig. 7.1.

Materials which do not meet these requirements may still have satisfactory resistance to SSC but this needs to be assessed by appropriate testing according to Section 8.2.4.

8.2.1. Materials Requirements

SSC can be prevented by limiting the strength, thereby eliminating deleterious crack-sensitive microstructures. Measuring the bulk hardness is a convenient method to check the strength of OCTG and other non-welded components, having homogeneous microstructures. It can be performed by various methods including Vickers HV30, Brinell and Rockwell, as detailed in Annex D.

In weld zones, which represent sites of possible tensile residual stresses and microstructural inhomogeneities, peak hardness and local hard zones are more important than bulk hardness. Hence Vickers HV5 and HV10 or Rockwell 15N are the preferred reference methods for hardness testing across welds.

The hardness at which a steel is resistant to SSC is dependent upon the severity of the environment. Therefore, the guidelines for allowable hardness levels may be relaxed in mildly sour environments provided appropriate testing demonstrates resistance to SSC.

Oil Country Tubular Goods (OCTG) and Non-welded Components

General guidelines:

- Quenched and tempered (Q + T) products are preferred for sour service,

Table 8.1. *Acceptable Vickers hardness (HV10) limits for welds*

Hardness Location	Region 3 Sour Service Domain	Region 2 Transition Domain	Region 1 $pH_2S > 0.0034$ bar $pH < 6.5$
Weld Metal and Heat Affected Zone:			
Root and Fill	250	280	300
Cap (wall < 9.5 mm thick)	275	300	325
Cap (wall ≥ 9.5 mm thick)	300	325	350

- Control the hardness to values ≤ 250 HV30 (≈ 22 HRC). (Q + T grades API 5CT C90 and T95 may be used at up to 270 HV30),
- Make sure that the hardness distribution across the wall (following the requirements of API 5CT) is uniform within 10% of the maximum value,
- Make sure that the microstructure is homogeneous and free from regions of untempered martensite.

API 5CT N80Q and C-95 can be used if continuously exposed to service temperatures in excess of 65°C. Above 80°C, API 5CT, N-80 and P-110 may be used, and above 107°C, API 5CT Q-125 may be used.[10]

Quenched and tempered components made of low alloy steels in the Cr, Mo series (AISI 4130 and 4140) are acceptable up to maximum hardness of 275 HV30 (26 HRC), although it is advisable to check the SSC performance of these alloys (Appendix A) when heat treated to hardnesses above 250 HV30 since microstructural variations may strongly influence sulphide stress cracking resistance.[11] If the components are cold straightened at or below 510°C, they must be stress relieved above 480°C.

Piping, Pressure Vessels and other Welded Components

The hardness of the parent metal should not exceed 250 HV30. In welds in these components distinction is made between the values of hardness of the inner and outer surfaces, i.e. hydrogen entry and hydrogen exit side, respectively. Acceptable peak hardness values are shown in Table 8.1. The acceptance of higher hardness in the weld and HAZ of the final cap pass only at the outer surface, which is generally not exposed to wet-H_2S, reflects the less severe hydrogen concentration at the hydrogen exit side.[1,2] In welding procedures where a final capping pass is made internally (e.g. in pressure vessels), the maximum acceptable hardness at this location would be as for the weld root.

When internal weld repairs or capping passes which have not been heat treated after welding are exposed to wet H_2S it is advisable to check the SSC-resistance of the as-welded structure as per Annex A.

If a single HV5 or HV10 measurement exceeds the acceptance criterion (by less than 25 HV) then four further measurements should be made adjacent to the four sides of this indentation at the prescribed minimum distance between adjacent indentations set out in ISO 65071-3. If these additional hardness values are below the acceptance criterion the material is accepted.

Note: Ni-containing weld consumables, as used to improve impact toughness values of the deposited weld metal have been found to have no ill-effect on weld metal SSC-behaviour up to at least 2.2% Ni content[12] provided hardness is controlled in accordance with this standard.

8.2.2. Cold Deformation Requirements

Cold forging or cold working during fabrication may render carbon and low alloy steels susceptible to SSC at hardnesses significantly below 250 HV. For any piping component which is cold-deformed to a permanent outer fibre deformation greater than 5%, heat treatment by normalising and tempering, or quenching and tempering is recommended. The SSC resistance of cold worked steels can be restored by thermal stress relief at temperatures in excess of 620°C although 650°C is necessary for microalloyed steels; low temperature stress relief at 590°C will not ensure SSC resistance. However, this requirement does not apply to pipe which is cold worked by pressure testing according to code requirements.

Tubulars and tubular components with permanent outer fibre deformation up to at least 8% (specifically pin nosing and/or box expanding of tubular ends prior to threading) with hardnesses up to 285 HV30 have been proven to be acceptable after stress relieving above a minimum temperature of 590°C. Care is required in the selection of stress relieving temperature as some tubular chemistries/heat treatments have required stress relieving temperatures of up to 650°C to return the SSC resistance back to the original pipe body level.

8.2.3. Free Machining Steels

Free machining steels may be susceptible to SSC at hardnesses as low as 160 BHN. Free machining steels are not recommended for sour service.

8.2.4. Qualification Tests

The SSC resistance of materials not complying with 8.2.1 and 8.2.2 can be assessed using the test methodology described in Annex A which presents general guidelines for selection of test method(s), test solution(s) and acceptance criteria. The test methods described may be used for the qualification of materials for use in a particular application. They may also be used to assess the SSC resistance of steels already in service in cases where the service environment has changed, e.g. by increased H_2S production. These tests are not intended for purchase order release purposes.

It is realised that some materials complying with 8.2.1 and 8.2.2 could have difficulties in fulfilling the proposed acceptance criteria. Such inconsistencies may be solved by detailed fitness for purpose studies (outside the scope of this document) using testing in simulated service environments at stress / strain conditions representative of in-service loads. In some respects laboratory tests may be more severe than service since there is all-side exposure to the environment and in some cases the relatively thin section of test pieces may result in a higher hydrogen concentration.

8.3. Guidelines to Avoid SWC

8.3.1. General

This section covers materials requirements for control of stepwise cracking. The recommendations in this section apply when the primary corrosion control method is by controlling the material properties. It is recognised that the control of corrosion by the use of coatings, cladding and inhibitors may in some operational circumstances be adequate engineering solutions to a potential SWC problem. Steel products intended for service in H_2S-containing environments may be tested in accordance with Annex B to evaluate their resistance to SWC.

Meeting the acceptance criteria given in Annex B should provide resistance to SWC in the majority of sour service applications. However, materials that fail to meet these criteria may still be suitable for use in mildly sour environments. In such cases, resistance to SWC may be demonstrated by conducting exposure tests in "real case" environments and adopting an acceptance criteria of "no cracking" or by performing large scale tests.

8.3.2. Seamless Pipes, Castings and Forgings

The susceptibility of seamless pipe to SWC is much less than of welded pipe,[13] however, field failures of seamless pipe have occurred because of SWC.[5, 14] For conventional, hot rolled, seamless products, a moderate limitation in sulphur content is all that is considered necessary to prevent severe SWC. A typical sulphur limit would be 0.01% S max. It should also be noted that seamless pipes are prone to blistering at the internal surface where the inclusions tend to be segregated. While this hardly affects the integrity of the pipe, it may interfere with monitoring of the wall thickness.

No special requirements are required for castings.

For conventional forgings the sulphur content should be limited to 0.025% maximum.

8.3.3. Rolled Steel

Rolled steel can be made with adequate SWC resistance provided that its chemistry and processing are controlled during manufacturing. The presence of inclusions (particularly elongated manganese sulphides) and bands of segregation and pearlite in the microstructure tend to reduce the resistance of a steel to SWC. Hence optimising resistance to SWC may involve the following:

Limiting sulphur content; low sulphur contents reduce the inclusion content and increase resistance to SWC.

Shape controlling inclusions; inclusions may be spheroidised using calcium treatment or rare earth treatment.

Minimising segregation; plates with heavy centreline segregation of carbon and manganese can be highly susceptible to SWC and should be avoided.

Minimising carbon, manganese and phosphorus as far as consistent with mechanical properties requirements; and

Controlling the plate rolling to avoid pearlite bands.

Steel materials for critical components (pipelines, pressure vessels, etc.) intended for H_2S service should be tested in accordance with Annex B.

8.4. Guidelines to avoid SOHIC and SZC

Measures to avoid SWC as described in Section 8.2. are also effective in reducing the risk of SOHIC and SZC.

Post-Welding Heat Treatment (PWHT), as applied to carbon and carbon-manganese steels for pressure vessels and piping reduces residual stresses and hardness differences across weld zones, thereby also significantly reducing the risk of SOHIC.

Resistance to SOHIC and SZC may be assessed by laboratory testing. In cases of high susceptibility it may be possible observe SOHIC in 4-point bend test samples or uniaxial tensile tests; however, large scale testing simulating the anticipated environmental conditions and mechanical stressing expected in service is also recommended. Such specialist testing is outside the scope of this document.

ANNEX A

Procedures and Guidelines for Sulphide Stress Cracking Testing

A.1. Scope

The content of this section describes the following test methods which are considered valid to evaluate the SSC-behaviour of materials.

Method A : Uniaxial tensile testing (UT) (smooth tensile specimen),
Method B : Four point bend testing (FPB),
Method C : C-ring testing (CR),
Method D : Double cantilever beam testing (DCB),
Method E : Slow strain rate testing (SSR).

A.2. Applicable Test Methods

The test methods are not equivalent and each can have a specific role. Method A indicates the fitness for service of a specific material in the given environment (see Appendix 1). Methods B and C are principally used to rank the performance of several materials in a given environment (see Appendix 2). Method D is applicable where a fracture mechanics design approach is used (see Appendix 3). Method E is useful for screening the performance of several materials as it is a quick means of assessment (see Appendix 4).

In the majority of carbon and low alloy steel products such as tubing, casing, piping and pressure vessels, the design code is based on a yielding criterion and the in-service-stresses are mainly tangential or axial, originating from internal pressure, end loads and weight. In these cases, the use of uniaxial testing, and/or C-ring testing, is adequate.

Further advice on selection of appropriate test methods is given in Table A.9 (Section A.9).

A.3. Test Solutions

The base test solution shall consist of 50 gL^{-1} NaCl + 4 gL^{-1} sodium acetate (CH_3COONa) in distilled or deionised water. This solution simulates the majority of produced waters and the buffering effect enables a wide range of solution pH values to be maintained. For tests requiring greater pH stability an alternative solution (NACE TM0177-Solution B) may be more appropriate. For example, this solution would be recommended for repeated routine testing. Where this solution is selected it shall be identified as "NACE TM0177-Solution B" quoting the adjusted test pH.

pH is adjusted to the selected test value by addition of HCl or NaOH. All pH values must be measured after saturation of the solution with the gas in the test.

If no specific data are available on the production environment the following pH conditions may be considered to be representative of typical production conditions:

A Solutions representing condensed water under CO_2 and H_2S pressure, as in gas producing wells, will use the test solution adjusted to pH 3.5.

B Solutions representing formation water under CO_2 and H_2S pressure, as in oil producing wells, will use the test solution adjusted to pH 4.5.

Saturation with gas containing mixtures of H_2S and CO_2 enables testing to be conducted at different H_2S partial pressure in the range 0.001 to 1 bar to simulate service environments of interest.

During the test, the pH may alter, but should not be allowed to change by more than 0.1 pH units. This should be achieved by periodically regenerating the buffering power of the test solution by pH adjustment. In addition, exclusion of oxygen from the test and maintenance of a sufficient solution volume to specimen area ratio according to the requirements of NACE TM0177 must be secured.

A.4. Test Temperature

The temperature shall be maintained at ambient (23°C ± 2°C) as this is the "worst case" expected in service for the occurrence of SSC in carbon and low alloy steels. For materials which will be *continuously* exposed to higher temperatures, higher test temperatures may be considered.

A.5. Reagents

These shall be in accordance with the requirements specified in NACE TM0177 and ISO 7539-1.

A.6. Acidic Gases

For both solutions A and B, Saturation is achieved by high purity H_2S gas (99.5% minimum purity). This will provide the most severe environment. However, for conditions requiring a less stringent environment to represent lower concentrations of H_2S, the solution can be saturated with a gas containing a mixture of H_2S and CO_2, enabling the use of different H_2S partial pressures in the range 0.001-1 bar. Buffering should be accomplished using sodium acetate additions. The H_2S content of the test solution should be checked by an appropriate method.

A.7. Specimen Geometry

All parent metal test specimens (with the exception of C-rings) shall be machined from the pipe or plate wall in the longitudinal direction. Crack progression in cylindrical tensile specimens is normal to the pipe axis, whereas in the double cantilever beam (DCB) specimen, cracking propagates parallel to the pipe axis. Appropriate choice of different specimen configurations will ensure that the cracking behaviour in both directions is studied.

In welds, tensile and four point bend specimens should be taken transverse to the weld, where feasible.

A.8. Test Vessels and Solution Volume

These shall be in accordance with the requirements specified in NACE TM0177 and ISO 7539-1.

A.9. Suggested Acceptance Criteria for the Various Test Methods

Table A.9 gives guidelines on proposed acceptance criteria depending upon the type of equipment being considered and the selected test purpose and method.

***Table A.9** Acceptance criteria for SSC test methods*

Type of Equipment	Purpose of Test	Preferred Test Methods	Suggested Acceptance Criteria[1]
Tubing, Casing Non-Welded Components (Thin Wall), $T/D \leq 0.05$	Fitness for Service	UT	$\sigma_{th} > 90\%$ Actual YS[3]
	Ranking	FPB	$\sigma_{th} > 90\%$ Actual YS[3]
		CR	$\sigma_{th} > 90\%$ Actual YS[3]
	Screening[2]	SSR	
Welded Piping, Pipeline and Pressure Vessels	Fitness for Service	UT[4]	$\sigma_{th} \geq 90\%$ Actual YS[3]
		FPB[4]	$\sigma_{th} \geq 90\%$ Actual YS[3]
		CR	$\sigma_{th} \geq 90\%$ Actual YS[3]
	Screening[2]	SSR	
Low-Pressure Containing Equipment	Fitness for Service	UT	$\sigma_{th} \geq$ Actual Service Stress
	Ranking	FPB	$\sigma_{th} \geq$ Actual Service Stress
		CR	$\sigma_{th} \geq$ Actual Service Stress
	Screening[2]	SSR	
Items Utilising Fracture Mechanics Design Basis[5]	Fitness for Service	UT	$\sigma_{th} \geq$ Actual Service Stress
	Design Using Fracture Mechanics	DCB[6]	35 MPa $m^{1/2}$
	Screening[2]	SSR	

1 More conservative acceptance criteria may be necessary for some applications.
2 Slow strain rate effects in service due to pressure variations or ripple loading due to vibrations, etc. may also be considered in some specific circumstances.
Tubing, casing etc., which occasionally may see load variations during work-over, is not considered to be in dynamic service.
3 Actual yield strength of material in finished form.
4 Specimens taken transverse to welds.
5 See Ref. [15].
6 Lower values of stress intensity may be applicable for high strength steels and acceptance criteria may be design dependent.

Appendix 1

Preparation and Use of Smooth Uniaxial Tensile Test Specimens (SSC Test Method A)

1.1 Method

Smooth specimen tensile testing shall be carried out in accordance with the procedure specified in NACE TM0177 (Method A). The additional requirements/ modifications are described in this Section.

Many types of stress fixtures and test containers used for stress corrosion testing are acceptable for the smooth specimen tensile test. These generally fall into two groups, namely constant load devices or sustained load (proof ring or spring loaded) devices (ASTM G49). For sustained load, the stiffness of the stressing device must be close to that of the sample.

1.2. Test Time

A period of 720 h is sufficient to evaluate the resistance to SSC. Prolonging the test may cause pitting and grooving at the surface which may give a false test result by providing stress raisers and local sources of hydrogen.

1.3. Applied Stress

SSC testing specimens have often been subjected to an initial load based on SMYS (e.g. 80% SMYS) in a severe test environment (1 bar H_2S, acidified). The philosophy developed within this document is to evaluate material performance under realistic conditions, in terms of both environment and stress level. It is considered for most major components (tubing, piping) that the residual stresses from production, welding, and laying will result in regions of the components being loaded near to or at actual yield. Hence, in principle, empirical testing should be conducted at actual yield. With provision for experimental errors and in order not to test above yield it is recommended to use an applied load of 90% of actual yield, and to check that the net stress of failed samples has not exceeded the yield due to cross section reduction by corrosion.

1.4. Specimens

Tensile specimens should be in accordance with NACE TM0177. If sub-size specimens are used, tensile specimens with 2.5 mm gauge diameter over a 25 mm gauge length shall be used, with a surface roughness $R_a \leq 0.2$ µm.

1.5. Failure Appraisal

Failure shall be considered as total fracture, or evidence of fissures or cracks in the specimens as determined by dye penetrant inspection where necessary.

When using a sustained load device, the gauge diameter of the test piece subsequent to exposure shall be free from any signs of fissures and / or cracks. This may be assessed through any of the following procedures:

(i) Visual examination of the sample using a binocular microscope of at least $\times 10$ magnification.

(ii) Metallographic examination of the gauge diameter by longitudinal sectioning and polishing.

(iii) Fracturing of the specimen using a tensile test machine at room temperature and subsequent analysis of the data and examination of the sample for signs embrittlement or brittle fracture.

Appendix 2

Preparation and Use of Four Point Bend Test and C-Ring Test Specimens (SSC Test Methods B and C)

2.1. General

The four point bend test may be used for SSC testing but is also appropriate for SOHIC evaluation. C-ring testing is particularly appropriate for evaluation of SSC-resistance of tubular components.

2.2. Method

Bent beam specimens should be prepared as described in ASTM G39 or ISO 7539 Part 2. The height and width dimensions of any specimen should not vary by more than 0.15 mm across the parallel faces for unwelded parent material. For welded joints, 4-point bend samples are preferred normally taken transverse to the weld and with the weld bead located at the centre of the sample (Figure A.2.1). Reduced thickness samples are permitted, provided that the change in the weld surface condition resulting from machining is recognised. For SOHIC evaluation the thickness shall be 15 mm or full wall-thickness if < 15 mm.

C-ring specimens should be prepared as described in ASTM G38 or ISO 7539 Part 5. The weld may be circumferential or axial relative to the C-ring (Figure A.2.2).

2.3. Applied Deflection

Following the situation in practice, no allowance is normally made for any stress concentration due to the weld profile in calculating the deflection necessary to obtain the required stress level for samples to be tested in the as-welded condition. Strain gauges can be attached to the side of the protruding weld bead on the test sample, or alternatively to a dummy sample, to determine the local stress corresponding to a given deflection.

It is permissible to derive the necessary sample deflection by calculation for both bent beam and C-ring samples, provided that the calculation procedure is validated against strain gauge measurements made on the steel being tested.

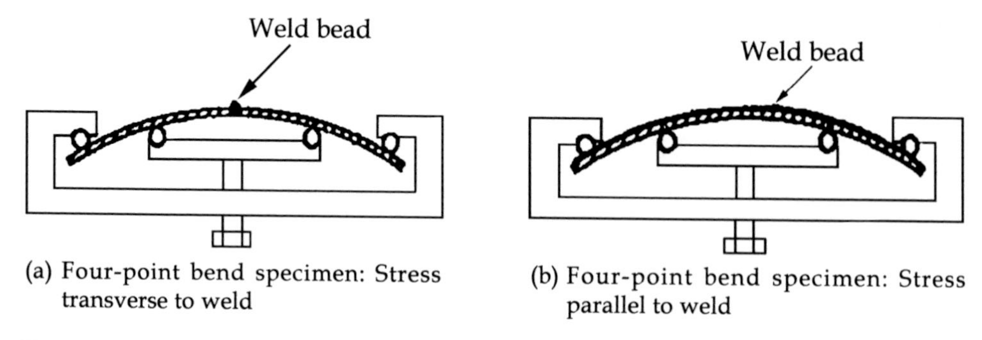

(a) Four-point bend specimen: Stress transverse to weld

(b) Four-point bend specimen: Stress parallel to weld

Fig. A.2.1 Schematic illustration of 4-point bend test rigs and samples.

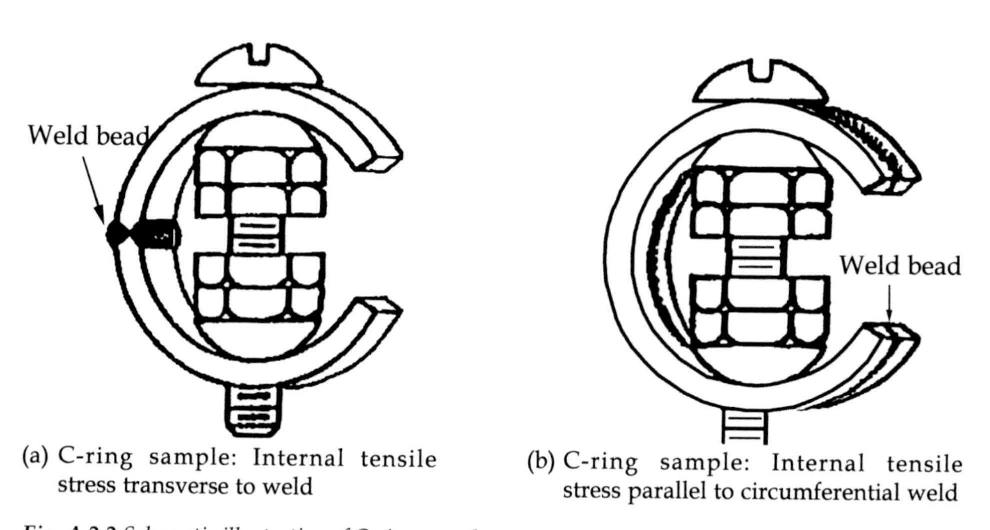

(a) C-ring sample: Internal tensile stress transverse to weld

(b) C-ring sample: Internal tensile stress parallel to circumferential weld

Fig. A.2.2 Schematic illustration of C-ring samples.

2.4. Failure Appraisal

Test specimens should be evaluated for any evidence of cracking including:

i) Surface breaking SSC.
ii) Sub-surface / surface breaking SOHIC and SZC.

The following evaluation method should be used:

i) Initial visual examination at 10 × magnification
ii) MPI of stressed test face
iii) Sectioning of the specimens at any suspicious features noted in steps (i) and (ii), otherwise at mid-width.
iv) Metallographic examination in the unetched condition at 100 × magnification of both cut faces. The size and location of any cracks should be confirmed in the etched condition.

All cracks identified should be reported, clearly identifying the type of crack and location.

Any stress corrosion cracks or SOHIC extending more than 0.1 mm in the through-thickness direction should be considered as cause for failure.

Any pits or other notable corrosion features should be recorded.

2.5. Reporting Requirements for Welded Specimens

When testing welded samples the following should be reported in addition to the requirements stated in ISO 7539 for parent materials:

(i) parent material and consumable types;
(ii) details of the welding procedure and welding parameters;
(iii) for C-ring and bent beam samples, the method of determining the deflection applied to attain the required stress level; and
(iv) weld orientation relative to principal stress.

Appendix 3

Preparation and Use of Pre-Cracked Double Cantilever Beam (DCB) Test Specimens (SSC Test Method D)

3.1. General

The use of DCBs is suggested for heavy sections/complex shaped items where fracture mechanics principles are meaningful.

Pre-cracked specimen testing provides qualitative as well as quantitative information. The latter is gained through a fracture mechanics approach. The data lead to the identification of a threshold stress intensity (K_{1SSC}) value for a material/ environment system. K_{1SSC} can be used as a fitness for service parameter when significant defects are expected and a fracture mechanics approach to design and component integrity is adopted. The values obtained in Test Method D are specific to the testing conditions and exposure time and provide primarily a ranking of tested materials. They should only be used for fitness for service purposes if the test conditions are similar to or more severe than those expected in service and if there is sufficient confidence in crack arrest during the test period.

3.2. Method

To identify a threshold stress intensity factor (K_{1SSC}), double cantilever beam (DCB) specimens shall be employed. Test and specimen geometry shall be in accordance with the procedures specified in NACE TM0177 (Method D). The additional requirements and modifications are described in this section.

3.3. Fatigue Pre-Cracking

All specimens shall be fatigue pre-cracked to at least 1.5 mm beyond the root of the chevron notch following the procedures described in NACE TM0177 and incorporating ISO 7539-6 recommendations.

3.4. Specimen Loading

DCB specimen loading shall be carried out either using a loading wedge (NACE TM0177: Method D) or by means of a screw placed through one arm.

3.5. Test Duration

The test duration shall be 15 days. The duration can be reduced if the potential drop technique is used to monitor the crack propagation and arrest. Testing for a longer period may be needed to achieve the correct K_{1SSC} value for certain material-environment combinations.

3.6. Failure Appraisal

The crack tip stress intensity shall be calculated from the relationship specified in NACE TM0177 and shall include a factor for the effect of the side grooves. In defining the K_{1SSC}, calculations shall be made of the crack tip stress intensity so that, where the crack comes to a halt, plane strain criteria are satisfied, i.e.

$$B \geq 2.5\ (K_{1SSC}/\sigma_y)^2$$

where σ_y = the yield strength (MPa) and B = specimen thickness (m).

The K_{1SSC} value shall exceed the acceptance criteria defined for the design or based upon the general guideline in Table A.9.

Appendix 4

Preparation and Use of Smooth Slow Strain Rate (SSR) Tensile Test Specimens (SSC Test Method E)

4.1. General

Slow strain rate tensile tests are versatile and adaptable for testing a wide range of product forms. The principal advantage of the test is the rapidity with which susceptibility to corrosion cracking can be assessed. It can be used as a rapid and efficient method of screening materials with respect to their resistance to SSC. Materials which seem to show reasonable SSC resistance can then be further evaluated using the techniques described in Appendices 1, 2 or 3.

Note: Materials can fail in slow strain rate testing that would never fail under field conditions. The test always produces a fractured specimen.

4.2. Method

A smooth specimen slow strain rate tensile test shall be carried out in accordance with the procedures specified in ISO 7539-7. The additional requirements and modification are described in this Section.

4.3. Test Apparatus

Specimens shall be pulled to failure using a stiff frame, slow strain rate tensile test machine.

4.4. Specimens

Specimens shall be as specified in Appendix 1, Section 1.4.

Note: To achieve consistency in results, the gauge length shall be fixed at 25 mm and the distance between the two shoulders shall be 30 mm for all specimens.

4.5. Extension Rate

The standard strain rate for this test method shall be 1×10^{-6} s^{-1}. This strain rate results in an overall extension rate of 2.5×10^{-5} mm s^{-1} for a specimen with 25 mm gauge length. This nominal strain rate represents an experimentally acceptable compromise; at higher speeds mechanical effects predominate and susceptibility to SSC may be missed while at lower speeds the time required for each test becomes excessive. Nonetheless, care should be exercised when deciding the extension rate.

4.6. Failure Appraisal

Evaluation of test data shall be carried out by means of assessing the normalised strain to failure (ε_n) and/or the embrittlement index (EI). These are described as follows:

Normalised Strain to Failure (ε_n)

Normalised strain to failure is defined as:

$$\varepsilon_n = \varepsilon_s / \varepsilon_{air}$$

where ε_s = strain to failure in the solution, and ε_{air} = strain to failure in air.

These values can be either the total strain to failure (elastic + plastic) or only the plastic strain to failure. In each case, consistency in data presentation shall be maintained. These properties shall be calculated from load-elongation curves which shall be recorded during each test.

Note: $\varepsilon_n = 1$ represents a fully resistant material (no susceptibility to SSC). Susceptibility to SSC increases as ε_n approaches 0.

Embrittlement Index (EI)

Embrittlement index shall be defined as:

$$EI = 1 - \frac{RA_s}{RA_{air}}$$

where RA_s = the reduction in area in the solution and RA_{air} = the reduction in area in air.

RA is the change in cross-sectional area of the test specimen. This is calculated as:

$$RA = 100\,(D_I^2 - D_F^2)/D_I^2$$

where D_I = initial gauge section diameter, and D_F = final gauge section diameter at fracture.

Note: EI = 0 represents a fully resistant material (no susceptibility to SSC) and EI = 1 represents total susceptibility to SSC.

ANNEX B

Procedures and Guidance on Test Methods for Stepwise Cracking

B.1 Scope

The content of this section should be used as a quality assurance measure to secure conformity in sour service performance of all delivered pipes and plates with respect to SWC. It is based on NACE TM0284 but with additional requirements and amendments.

B.2. Test Method

The test method complies with NACE TM0284 with the following amendments/ clarifications.

The four principal surfaces of each specimen shall be ground wet or dry and finished to a surface roughness R_a of 1 μm or finer which may be achieved using 320 grit. The two end faces of each test specimen are reserved for identification stamping and hence do not require the same surface preparation and surface finish as the four major faces.

Heavy wall pipes and plates with thickness above 30 mm can be tested in full wall thickness as one specimen as long as the width of the specimen remains 20 mm. However, if difficulties with surface preparation of such large specimens are envisaged, an odd number of subspecimens staggered across the wall having a convenient thickness of not less than 10 mm and an overlap of 1 ± 0.5 mm between adjacent subspecimens may be used. In using subspecimens, care should be exercised to avoid having inclusions or segregated zones (e.g. from the plate centreline) too close to the surface of the specimen as this will enhance the propensity for blister cracking at these sites which may not be representative of service performance.

After machining and immediately before commencing the test, the surfaces of the specimens shall be treated to remove any grease, inhibitors or other contaminants by metallographic grinding or by exposure to an oxidising acid.

Test specimens shall be placed in the nitrogen purged test tank before admission of predeareated and H_2S presaturated test solution into the tank.

B.3 Test Solution

The test solution shall be in accordance with NACE TM0284 (low pH, solution A).

The pH of the test solution shall be recorded immediately after start of test and also immediately before termination of the test. The concentration of H_2S in solution should be recorded at the same times and be $\geq$ 2000 ppm as determined by iodometric titration.

B.4 Test Temperature

The test temperature shall be $23 \pm 2°C$.

B.5. Number of Test Specimens

A number of sets of test specimens representative of the full range of production conditions shall be tested.

B.6. Position of Test Specimens

The positions of test specimens are as described in NACE TM0284. Where the pipe size is too small to allow the use of flat machined specimens, curved specimens may be used. Such specimens should not be flattened prior to testing.

B.7 Evaluation

By agreement prior to contract and start of production, the purchaser may specify evaluation by metallographic and/or ultrasonic methods.

Metallographic evaluation shall comply with NACE TM0284 with the additional requirement that all faces shall be subjected to either wet magnetic particle testing or macroetching before final metallographic polishing, in order to make an accurate assessment of whether any significant cracks are present which may have become invisible due to smearing of the metal surface during preparation.

The macroetching can be performed on the faces after grinding in steps from 80 grit to 500 grit; 10-15% ammonium peroxidisulphate etchant will reveal the crack clearly in a macro-examination.

Ultrasonic Evaluation of Coupons

Ultrasonic evaluation of coupons may be used by special agreement between manufacturer (or testing laboratory) and purchaser, as this will give increased information of the extent of planar cracks.

Any equipment utilising ultrasonic principles and capable of continuous inspection of the entire surface along one of the major axes, can be used. The equipment shall be checked with an applicable reference standard at least once per continuous lot of samples tested per day, to demonstrate its effectiveness and the inspection procedures.

The reference standard shall contain well defined flat bottom drilled holes.

The inspection procedure and calibration of the equipment shall be demonstrated, by the user, to yield consistent information on the position and extent of the defects. Also, the user shall provide a correlation curve between ultrasonic indication evaluation and metallographically determined crack length ratio or crack sensitivity ratio.

B.8 Acceptance Criteria

The values of crack length ratio (CLR), crack surface ratio (CSR) and crack thickness ratio (CTR) given below should be considered as target values only. Strict limits are difficult to justify as there is probably no difference in the service behaviour of steels with a wide range of CLR. In this respect "good" quality steel will normally have CLR values well below 20% whereas "poor" quality steels may have CLR values well above 50%.

The following acceptance criteria for materials tested in accordance with this Annex are recommended:

Crack length ratio, CLR $\leq$ 15%
Crack thickness ratio, CTR $\leq$ 5%
Crack surface ratio, CSR $\leq$ 2% (5% if all cracks are within the centre segregation zone and there is no crack face separation > 0.1 mm).

Note:

- All cracks with any part lying within 1 mm below the test surfaces are discarded from the calculation.
- If subdivided specimens staggered over the thickness of heavy wall materials are used, all cracks within each set of specimens covering the full wall thickness shall be considered to be cumulative and CLR, CSR and CTR values shall be evaluated as for one full size specimen.
- It should be noted that CTR and CSR are thickness dependent measures. Higher acceptance criteria may be appropriate for thin sections, e.g. <8 mm.
- The acceptance criteria for each test sample should be applied to the average of each ratio from the three cut faces examined.

ANNEX C

Guidelines for Determination of pH

Figures C.1–C.5 can be used as a quick guide to assess an approximate value of the pH of the water phase for various conditions if no accurate pH calculations or measurements are available. The deviation may be taken as +0.5 and −0.0 pH units.

Figure C.1 The pH of condensate waters under CO_2 and H_2S pressure.

Figure C.2 The pH of limestone unsaturated formation waters under CO_2 and H_2S pressure.

Figure C.3 The pH of limestone saturated or supersaturated formation waters under CO_2 and H_2S pressure at 20°C.

Figure C.4 The pH of limestone saturated or supersaturated formation waters under CO_2 and H_2S pressure at 60°C.

Figure C.5 The pH of limestone saturated or supersaturated formation waters under CO_2 and H_2S pressure at 100°C.

Note that actual pH values in highly concentrated chloride environments (>0.5 M NaCl) may be lower than the estimates indicated in these figures.

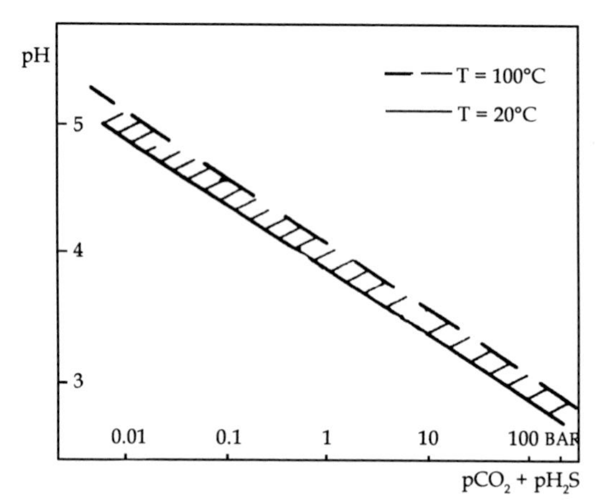

Fig. C.1. *The pH of condensate waters under CO_2 and H_2S pressure.*

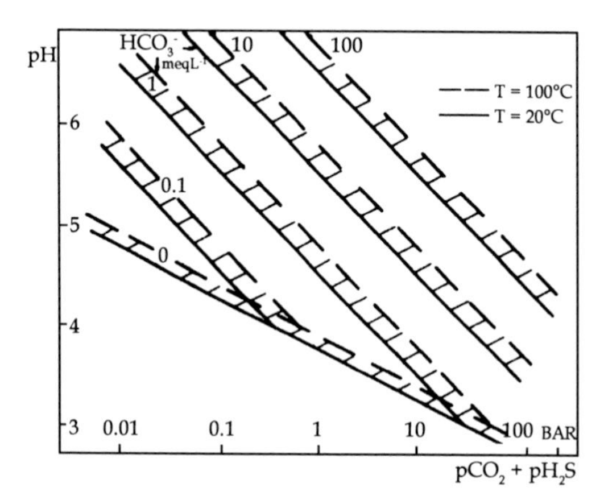

Fig. C.2. *The pH of limestone unsaturated formation waters under CO_2 and H_2S pressure.*

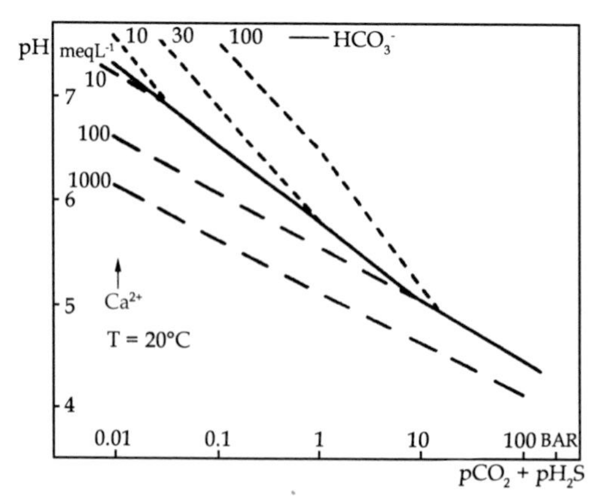

Fig. C.3. *The pH of limestone saturated or supersaturated formation waters under CO_2 and H_2S pressure at 20°C.*

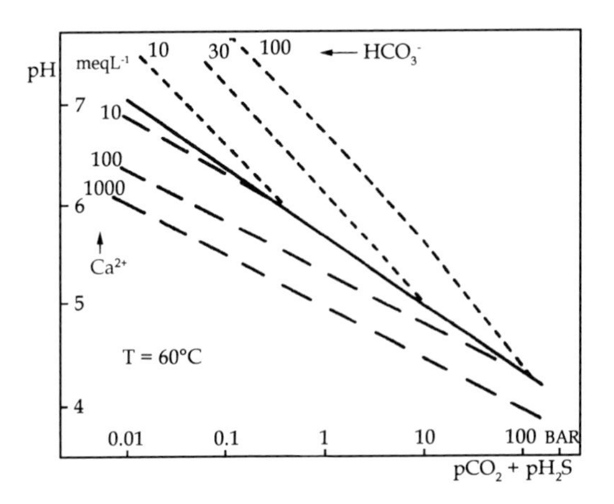

Fig. C.4. *The pH of limestone saturated or supersaturated formation waters under CO_2 and H_2S pressure at 60°C.*

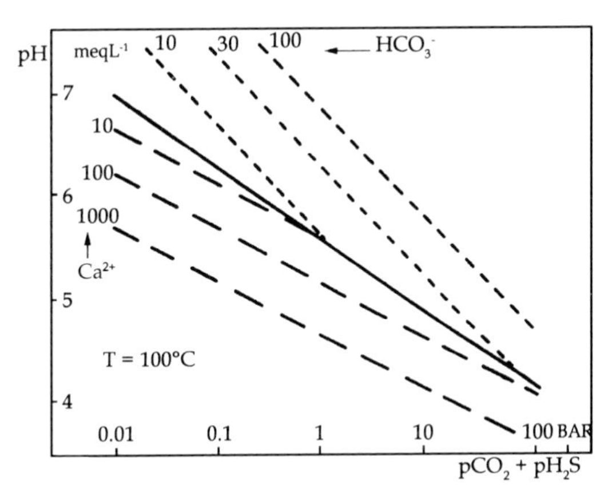

Fig. C.5. *The pH of limestone saturated or supersaturated formation waters under CO_2 and H_2S pressure at 100°C.*

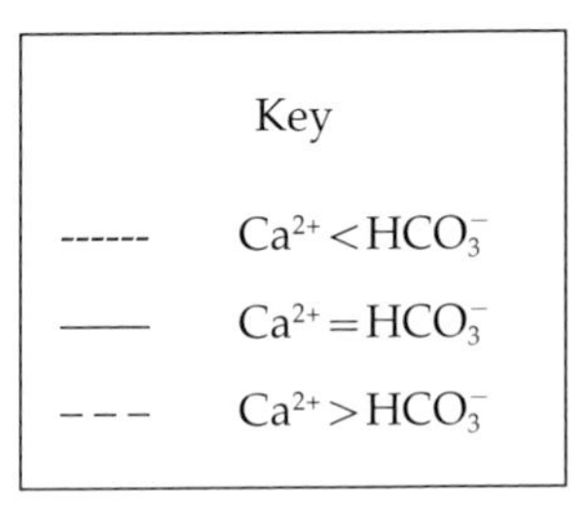

ANNEX D

Hardness Testing of Components and Weld Zones for Service in H_2S-Containing Environments

D.1. Scope

This Annex is intended to give further background information to Section 8.2.1 of this document which stipulates hardness limits for carbon and low alloy steels operating in sour service. It gives guidance on test techniques and protocols, which will ensure that the true hardness is assessed as accurately as possible Although it is recognised that cracking can also occur in soft microstructures, particularly, for example, in regions weld heat affected zones (HAZs), assessment of such regions is not covered by this document.

D.2. Significance of Welds

Consideration must be given to whether the material is welded or not. Oil country tubular goods, for example, are unwelded and, in general, are homogeneous in structure. Precise location of hardness measurement, and the amount of material sampled by each impression, are less important in such circumstances. Pipelines and pressure vessels, however, are necessarily welded, and thus there are inhomogeneities on a fine scale, which may contain hardened microstructures, particularly in the heat affected zone. Accurate assessment of such regions requires careful attention to the location and size of hardness impressions. Because welds are the sites of residual stresses and chemical inhomogeneities, as well as possible small hard regions, they are at particular risk from environmental cracking and accurate assessment of peak hardness is of paramount importance.

D.3. Hardness Testing Techniques

The principal hardness testing techniques are given in Table D.1.

Conversion factors between the different scales for homogeneous ferritic steels are published in BS 860 and ASTM E140. Comparison of the conversions given in the two standards for Vickers, Brinell and Rockwell C (which were presumably

Table D.1 *Common hardness testing techniques*

Name	National Standard	Load	Equivalent Hardness Values[1]	Diameter of Impression, mm
Vickers	BS 427	50 g	248	0.019
	ASTM E92	100 g	248	0.027
	ASTM E384	500 g	248	0.061
	(micro)	5 g	248	0.19
	DIN 50133	10 g	248	0.27
	ISO 6507 1-3	30 g	248	0.48
Rockwell HRC	BS 891	150 kg	22	0.80
	ASTM E18			
	DIN 50103			
	ISO 6508			
Rockwell HR15N	ASTM E18	15 kg	70.5	0.27
Brinell	BS 240	3000 kg	237	3.93
	ASTM E10	1500 kg	(237)	2.81
	DIN 50351			
	ISO 6506			
Knoop	ASTM E384	10 g	(258[2])	0.023
		50 g	(258[2])	0.053
		100 g	(258[2])	0.074
		500 g	258[2]	0.17
Brinell Comparator	ASTM A833	Impact	–	< 4.2
Equotip	–	11.0 N mm Impact	–	0.54[3]
Microdur	–	5 kg	248	0.19

[1] From ASTM E140.
Figures in parentheses are using a load not referred to in ASTM E140
[2] Extrapolation of data in standard. Long dimension
[3] Measured on a 254 HV5 standard hardness block.

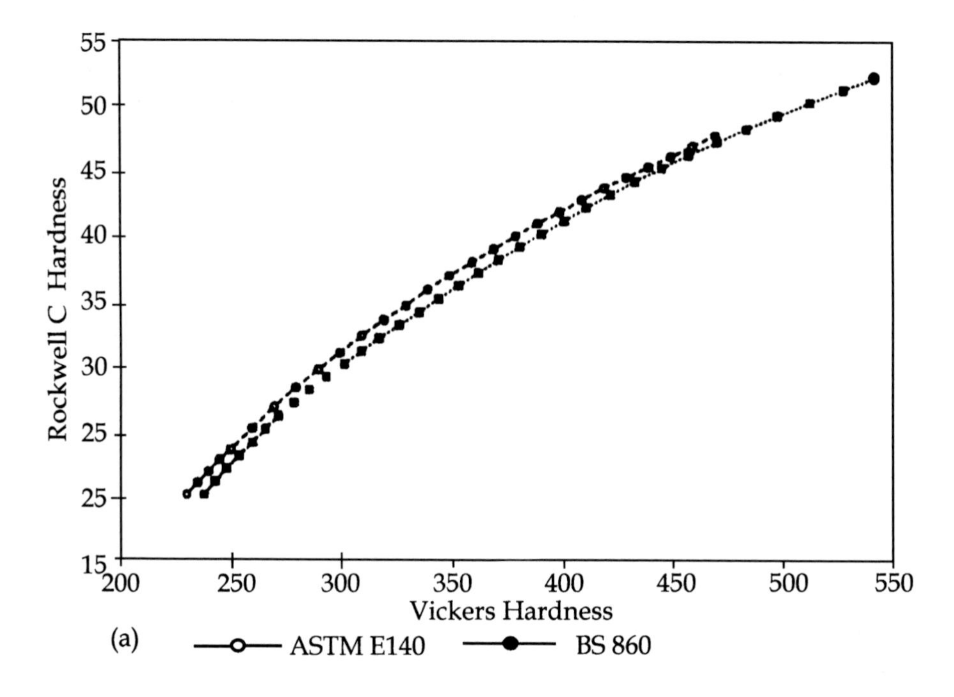

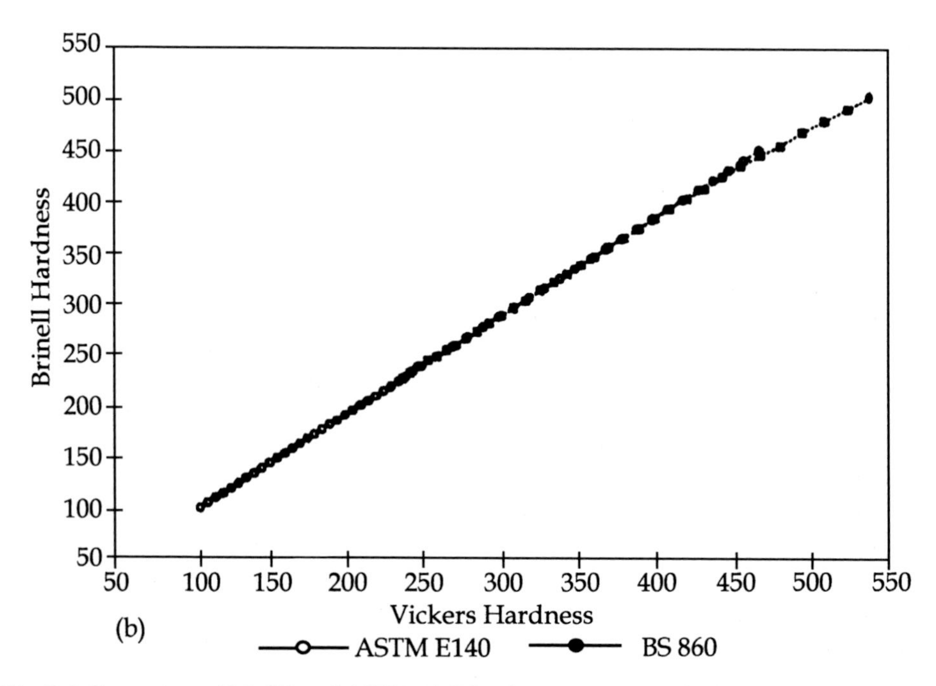

Fig. D.1 *Comparison of BS 860 and ASTM E140 hardness conversions for ferritic steel. (a) Vickers: Rockwell and (b) Vickers : Brinell.*

generated independently) shows that they are very close (Fig D.1), and gives some assurance of the reliability. For homogeneous materials conversions may also be based on testing using different hardness test techniques on the same block of material. Note that these conversions are, however, applicable only to homogeneous materials, and can seriously underestimate the hardness of heat affected zones.

From the above comments, it may be concluded that the most reliable way of assessing weld area hardness is to use a technique with as small an impression as possible. This is not necessarily the case, however, as the accuracy of the actual hardness measurement generally falls as the size of the impressions decreases. The recommended technique is testing in accordance with HV10 or HR15N. For particularly low arc energy welds (< 25 kJ cm^{-1}) HV5 testing is recommended.

Hardness testing using Equotip and Microdur methods are not appropriate for hollow components such as pipe.

D.4. Location of Hardness Impressions

For components which are not welded, and for which no variation in hardness is anticipated from other manufacturing operations (e.g. cutting, forming, or heat treatment), the precise location of hardness impressions is not important. It is advisable to ensure that measurements cover as large an area as conveniently possible, to give some assurance of homogeneity. Furthermore, it is more important to know the hardness of material which will be in contact with or close to the service environment, than material on the outside of a vessel or pipe.

For OCTG, the requirements of API 5CT should be followed for location of hardness impression.

For welded components, the hardness survey should be carried out in accordance with BS 4515 which describes the locations of hardness impressions. Ideally the survey should determine the maximum hardness present, concentrating on material which will be in contact with or close to the service environment. For ferritic steels, maximum hardness would normally be anticipated in the HAZ, close to the fusion boundary, or possibly in the weld metal. Peak hardness will commonly be found for low arc energy welds, and unfortunately these also give rise to narrow HAZs which are particularly difficult to sample accurately. Weld metal or HAZ which has seen little or no reheating by subsequent passes will in general be hardest. Hardness in a HAZ generally peaks very close to the fusion boundary although in a typical C-Mn steel weld, some carbon migration is possible from the generally higher carbon plate to the lower carbon weld metal, giving peak hardness just within the weld. (With modern very low carbon linepipe steels this effect should be reduced). Furthermore, as the material which affects a hardness reading extends beyond the impression itself, maximum hardness will not generally be recorded for impressions which actually touch the fusion boundaries. Measurements within about 0.4 mm of the section boundary can generally be relied upon to give an accurate measurement of HAZ hardness. Regions of high hardness are often identifiable as areas which etch heavily in a standard nital etch, but light etching regions should also be examined, as very hard martensitic regions can be resistant to etching.

In some situations, it may be necessary to carry out weldment hardness tests in situ. In making site hardness measurements, careful preparation of the test area is essential both to obtain most accurate results and to ensure that the correct weld region is sampled. However, even with such care, it must be recognised that site hardness testing can give misleading results, The hardness values given assumes that the method and indenting load are selected appropriate to the region of concern.

References

1. M.B. Kermani et al., Limits of Linepipe Weld Hardness for Domains of Sour Service in Oil and Gas Production, Corrosion 2000, Paper No.157, NACE, Houston, TX, U.S.A.
2. R. J. Pargeter, The Effect of Low H_2S Concentration on Welded Steels, Corrosion 2000, Paper No.143, NACE, Houston, TX, U.S.A.
3. EFC Publication 17, A Working Party Report on Corrosion Resistant Alloys for Oil and Gas Production: Guidance on General Requirements and Test Methods for H_2S Service, 2nd Edn. 2002, The Institute of Materials, London, U.K.
4. W. Bruckhoff et al., Rupture of a Sour Gas Line Due to Stress Oriented Hydrogen Induced Cracking, Corrosion '85, Paper No. 389, NACE, Houston, TX., USA.
5. M. G. Hay and M. D. Stead, The Hydrogen Induced Cracking Failure of a Seamless Sour Gas Pipeline, NACE Canada Region Western Conference, 7–10 February, 1994, Calgary, Alberta, Canada.
6. M. B. Kermani, D. Harrop, M. L. R. Truchon and J.-L. Crolet, Experimental Limits of Sour Service for Tubular Steels, Corrosion '91, Paper No. 21, NACE, Houston, TX, USA.
7. J-L. Crolet and M. R. Bonis, pH Measurements under High Pressures of CO_2 and H_2S, Mat. Perform., 1984, 23, 3542.
8. J-L. Crolet and M. R. Bonis, An Optimized Procedure for Corrosion Testing under CO_2 and H_2S Gas Pressure, Corrosion,1990, 46, 81–86.
9. M. Bonis and J-L. Crolet, Practical Aspects of the Influence of in situ pH on H_2S Induced Cracking, Corros. Sci., 1987, 27, 1059–70.
10. NACE standard MR0175, Materials for use in H_2S-containing Environments in Oil and Gas Production, NACE, Houston, TX, USA.
11. M. Watkins and R. Ayer, Microstructure - The Critical Variable Controlling the SSC Resistance of Low Alloy Steels, Corrosion '95, Paper No. 50, NACE, Houston, TX, USA.
12. T. G. Gooch, SCC of Ferritic Steel Weld Metal — The Effect of Nickel, Metal Construction, 1982, 14 (1), 29-33; 14 (2), 73–79.
13. D. Sourdillon, G. Guntz and A. Sulmont, Susceptibility of Seamless and Welded Line Pipe to HIC, Vallourec Research Centre, 1979
14. S. Y. Gayan and A. El-Amari, Failure of a Gas/Condensate Line, Mat. Perform., 1992, 31 (10), 55.
15. BS 7910:2005 Guide to methods for assessing the acceptability of flaws in metallic structures.